Addition

by Kristin Sterling

first step nonfiction

Lerner Publications Company · Minneapolis

I can add.

To add means to join together.

This girl has three cars.

This girl has two cars.

Together they have five cars.

3 + 2 = 5

Three plus two equals five.

$$2 + 2 = ?$$

Can you add?